Early
NATURAL DISASTERS
Encyclopedias

TORNADOES

by Meg Marquardt

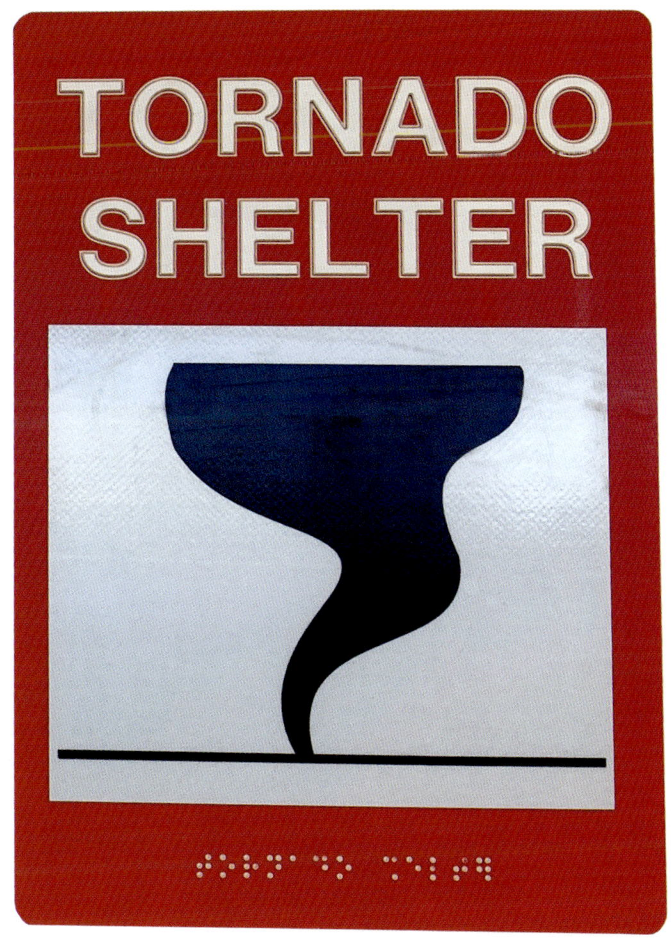

Early Encyclopedias

An Imprint of Abdo Reference
abdobooks.com

abdobooks.com

Published by Abdo Reference, a division of ABDO, PO Box 398166, Minneapolis, Minnesota 55439.

Printed in China.
102024
012025

Editor: Marie Pearson
Series Designers: Candice Keimig, Joshua Olson
Production Designer: Ryan Gale

Library of Congress Control Number: 2024938364

Publisher's Cataloging-in-Publication Data

Names: Marquardt, Meg, author.
Title: Tornadoes / by Meg Marquardt
Description: Minneapolis, Minnesota: Abdo Reference, 2025 | Series: Early natural disasters encyclopedias | Includes online resources and index.
Identifiers: ISBN 9781098296056 (lib. bdg.) | ISBN 9798384917052 (ebook)
Subjects: LCSH: Tornadoes--Juvenile literature. | Natural disasters--Juvenile literature. | Weather--Juvenile literature. | Severe storms--Juvenile literature. | Meteorology--Juvenile literature. | Earth science--Juvenile literature. | Encyclopedias and dictionaries--Juvenile literature.
Classification: DDC 363.34--dc23

CONTENTS

Not all thunderstorms form tornadoes.

A Stormy Night

The sun is hidden behind dark storm clouds. Lightning strikes. Hail hits the sidewalk. Suddenly, tornado sirens wail. A tornado is forming.

Deadly Winds

Tornadoes are spinning columns of air. They form in severe thunderstorms. Tornadoes stretch from clouds to the ground. They travel an average of 10 to 20 miles per hour (16 to 32 kmh). They can spin up to 300 miles per hour (480 kmh).

Tornadoes can kick up dust and objects near the ground.

The average tornado family has two to three tornadoes.

2013 El Reno Tornado

In 2013, the widest tornado ever recorded hit El Reno, Oklahoma. It was 2.6 miles (4.2 km) wide. Its wind speeds were nearly 300 miles per hour (480 kmh).

Many Types

Some tornadoes are long and skinny. Others look wider than they are tall. Sometimes a single storm has more than one tornado. These tornadoes are called a family.

Hard to Predict

Tornadoes are hard to predict. Only 1 percent of thunderstorms form tornadoes. Certain temperatures, wind speeds, and humidity are needed. Meteorologists are scientists who study weather. They work to better understand tornadoes.

Meteorologists use tools to track storms that could form tornadoes.

Tornado Hot Spot

The United States gets more tornadoes than any other country. About 1,200 tornadoes happen there every year. The features of the land help tornadoes form.

Colorado gets about 50 tornadoes each year.

Satellites can track storms.

Weather Systems

Weather systems are the movement of cold and warm air around Earth. Some systems travel from the Pacific Ocean over the Rocky Mountains. Others come down from Canada. More move up from the Gulf of Mexico. They all meet over the Great Plains.

The Great Plains

The Great Plains stretch from Texas up to Canada. They are mostly between the Mississippi River and the Rocky Mountains. The land is flat. When weather systems meet here, they can make thunderstorms. Then tornadoes can happen.

The Great Plains have few trees.

The unstable air in Tornado Alley makes tornadoes relatively common.

Tornado Alley

Tornado Alley is a nickname given to a part of the United States. Tornadoes often happen here. The area moves through the year. In cool months, it is in the Southeast. As temperatures rise, it shifts to the southern plains. By late summer, it is in the Midwest.

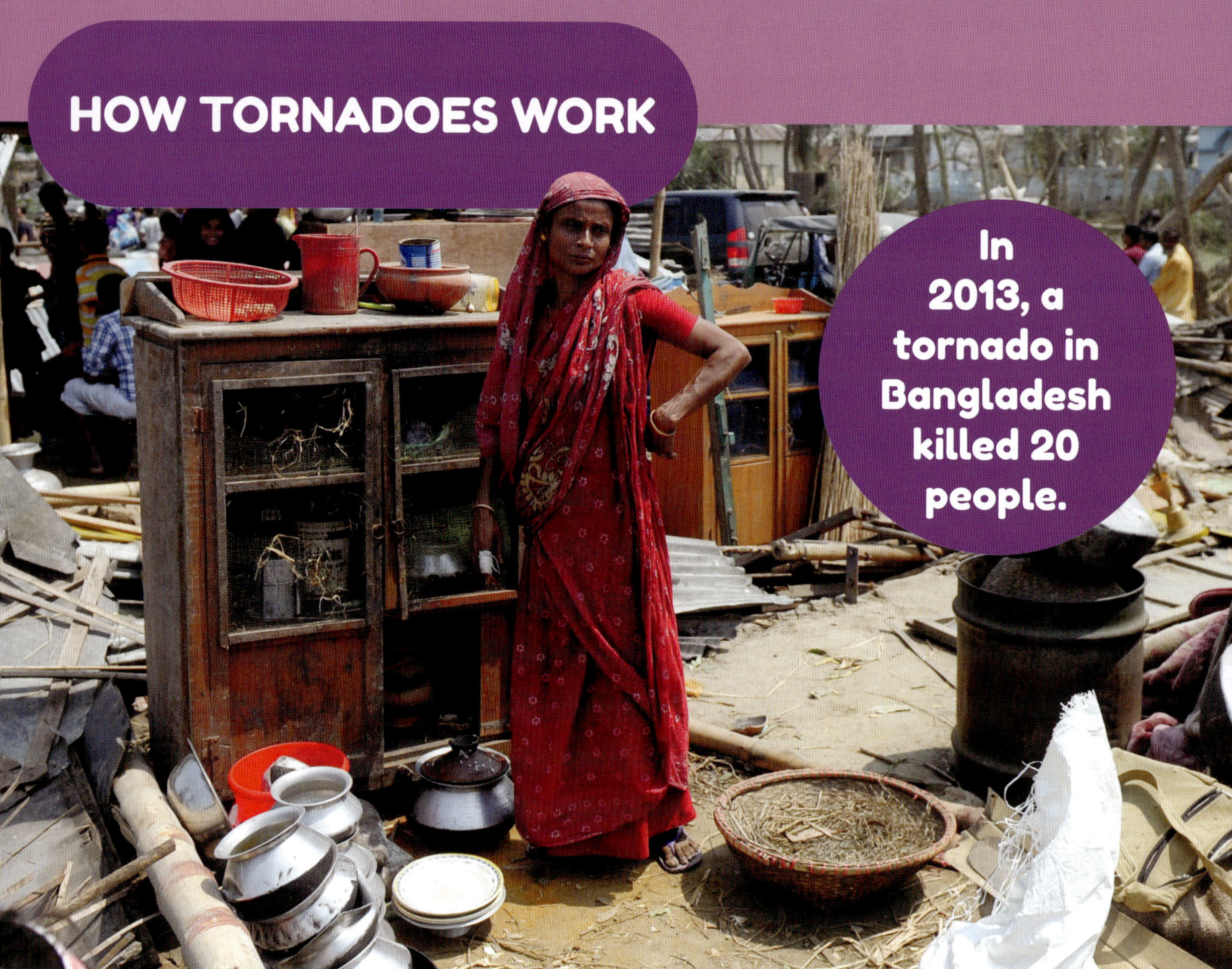

In 2013, a tornado in Bangladesh killed 20 people.

Deadly Seasons

Bangladesh is in Southeast Asia. It has one of the deadliest tornado seasons in the world. Cold air comes down from the Himalaya Mountains. Hot air comes up from the Bay of Bengal. These fronts meet over Bangladesh.

Weather Conditions

Tornadoes need several ingredients. The first is thunderstorms. Thunderstorms have lightning.

FUN FACT!

Tornadoes can happen in any season. But they are most common in spring and summer.

Lightning is a flash of electricity that travels through the air.

Wind Shear

Wind shear is another ingredient. Wind shear is when winds move in different ways. They could move at different speeds. Or they could move in different directions. This can cause wind rotation.

Wind shear can describe both horizontal and vertical wind movement.

Severe storms often happen in the evening. They need the heat from the day to rise into the air.

Air Type

Air type is also important. Sometimes air near the ground is warm and moist. In the storm clouds, the air is cool and dry. The warm air rises. It can change the moisture and temperature in the clouds. The change is called atmospheric instability. Tornadoes need it to form.

Waterspouts

Tornadoes can form over water. They are called waterspouts. Waterspouts can move onto land. If that happens, they are called tornadoes.

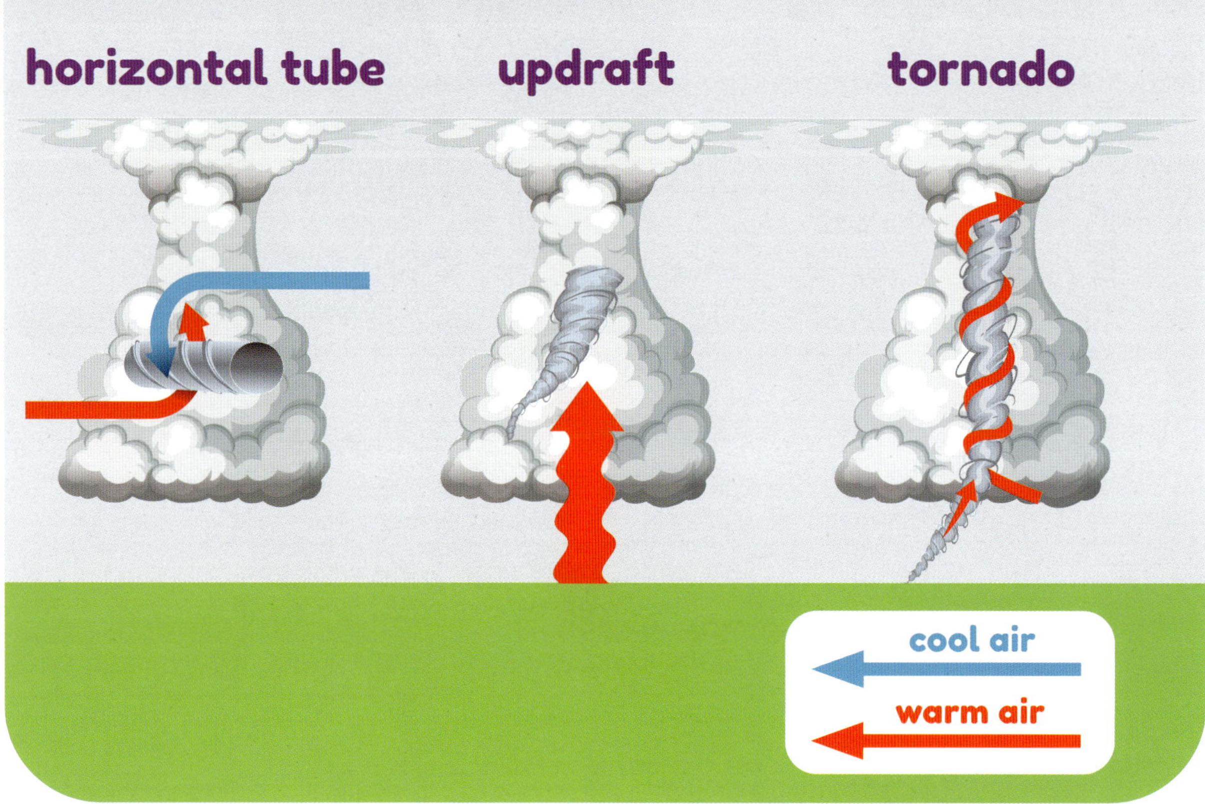

Wind shear causes horizontal tubes. These tubes are made of warm air. That warm air starts to rise in an updraft. Then a tornado can form.

Updraft

Wind shear and atmospheric instability mix together. They make an updraft. An updraft is air moving quickly upward. The updraft may spin.

Funnel Cloud

The spinning updraft has a tube shape. It collects cloud droplets. These make the tube visible. The visible tube is called a funnel cloud.

Rain Wrapping

Rain can fall so hard that it becomes hard to see very far. When a tornado forms in a lot of rain, it's not easy to spot. It is called a rain-wrapped tornado.

A funnel cloud shows only the inner core of the spinning air.

Touching Down

The funnel cloud becomes a tornado when it reaches the ground. On average, tornadoes stay on the ground for five to ten minutes. Some touch the ground for only a moment. Others last for hours.

Supercells are the rarest type of thunderstorm.

Supercell Thunderstorms

Tornadoes most often come from supercell thunderstorms. These huge storms can reach 50,000 feet (15,240 m) high. They can be ten miles (16 km) across.

FUN FACT!

Most tornadoes strike between 3:00 p.m. and 9:00 p.m.

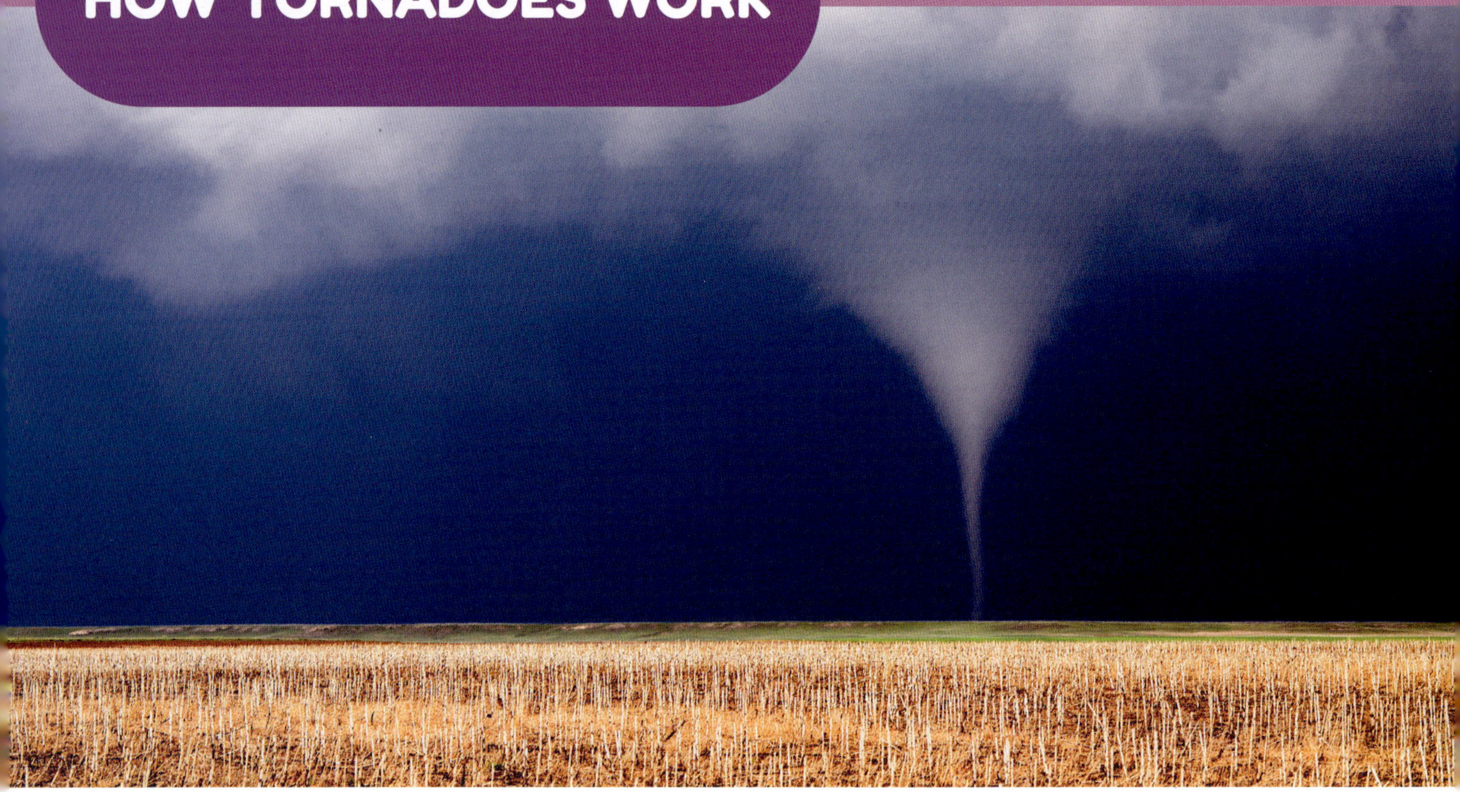

Weak tornadoes usually last just a few minutes.

Weaker Thunderstorms

A thunderstorm that lasts no more than an hour can also create a tornado. These storms usually form weaker tornadoes than supercells do. The tornadoes do not last as long.

FUN FACT!

Hurricanes can cause tornadoes.

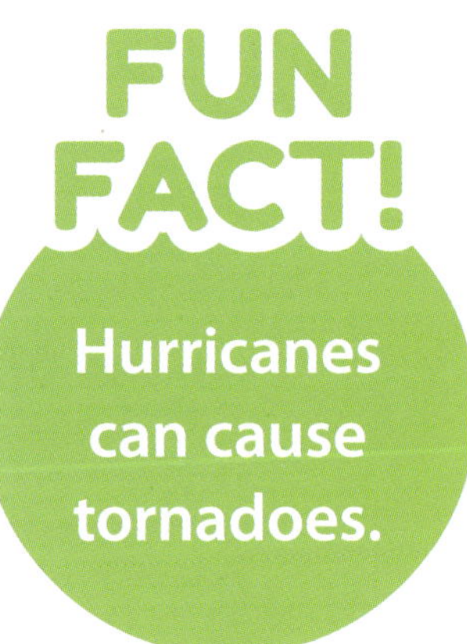

Measuring Strength

The Enhanced Fujita Scale (EF Scale) measures tornado strength. The scale goes from EF0 to EF5. An EF0 tornado has wind speeds between 65 and 85 miles per hour (105 and 137 kmh). An EF5 tornado can reach more than 200 miles per hour (320 kmh).

EF Scale

The Enhanced Fujita Scale has been in use since 2007.

EF Scale	Wind speed	Damage
0	65–85 miles per hour (105–137 kmh)	Some roof, gutter, and siding damage
1	86–110 miles per hour (138–177 kmh)	Severe roof stripping, mobile homes overturned, broken windows
2	111–135 miles per hour (179–217 kmh)	Roofs torn off, mobile homes destroyed, house foundations shifted
3	136–165 miles per hour (219–266 kmh)	Complete stories of houses destroyed, severe damage to large buildings
4	166–200 miles per hour (267–322 kmh)	Houses completely destroyed
5	Faster than 200 miles per hour (322 kmh)	Houses completely destroyed and blown away from foundations, significant damage to high-rise buildings.

Measuring Damage

Meteorologists estimate an EF Scale number based on damage. They might look at damaged homes. If only bits of roofs blew off, it was probably an EF0 tornado. Blowing off an entire roof takes stronger winds. That tornado was likely an EF2.

Weak tornadoes might not damage any part of a house but the roof. People cover leaking sections in tarp until the roof can be repaired.

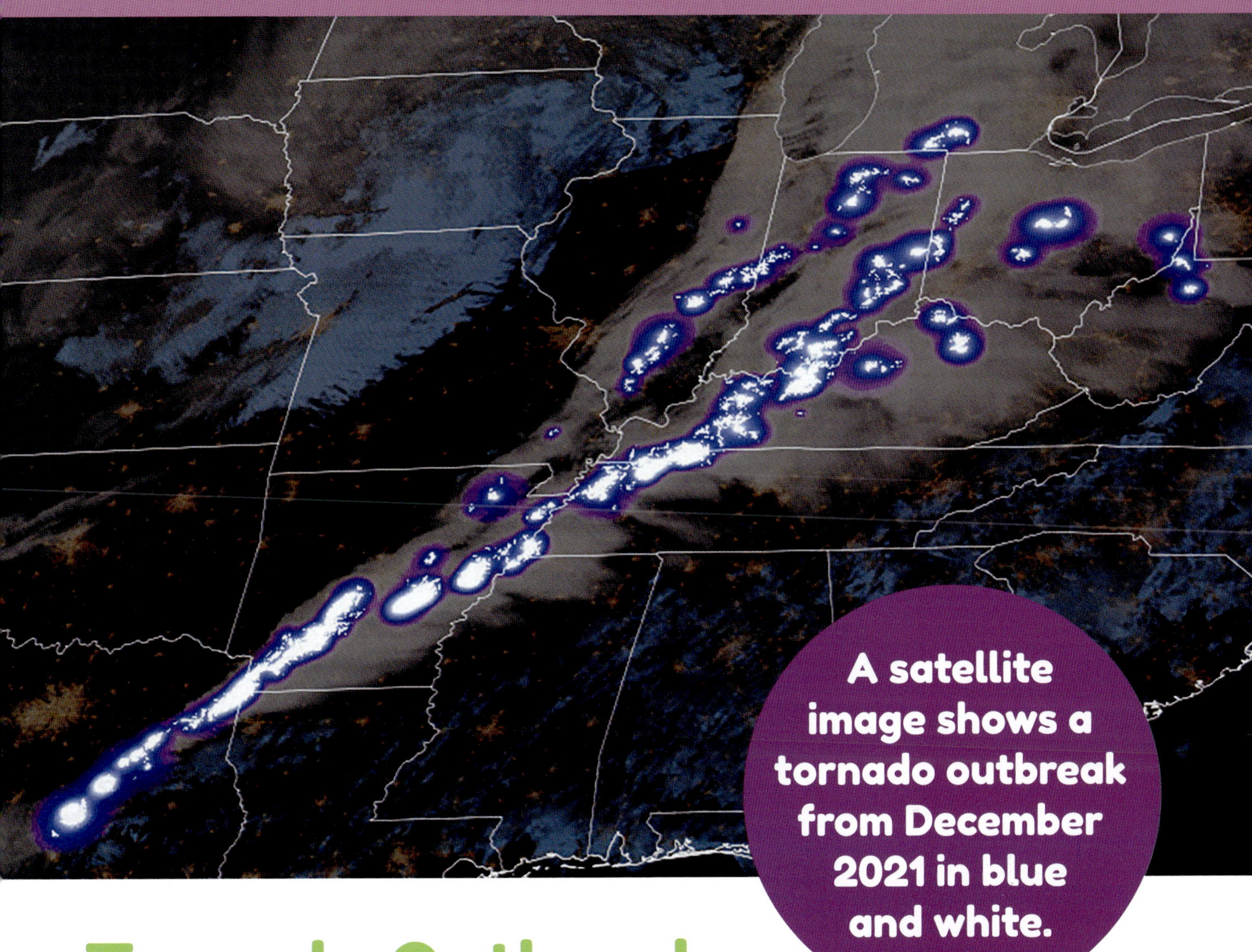

A satellite image shows a tornado outbreak from December 2021 in blue and white.

Tornado Outbreaks

A group of storms is called a system. Large storm systems can last many days. One system can have many tornadoes. This is called a tornado outbreak. The outbreaks often have long-track tornadoes.

An EF3 tornado hit Tennessee in December 2021 and traveled almost 123 miles (198 km).

Long-Track Tornadoes

A long-track tornado stays on the ground longer than most. *Track* refers to a tornado's path along the ground. Most tornado tracks are about 3.5 miles (5.6 km) long. Long-track tornadoes travel much farther. They usually go at least 15 miles (24 km). Some can travel more than 100 miles (160 km).

Multiple-Vortex Tornadoes

Sometimes a tornado has more than one funnel. This is called a multiple-vortex tornado. *Vortex* is another word for funnel. The funnels can rotate around each other. Or they can rotate around a center funnel.

FUN FACT!

Tornadoes usually spin counterclockwise in the Northern Hemisphere. They often spin clockwise in the Southern Hemisphere.

Sometimes the main funnel is accompanied by a weaker funnel.

Skipping

A tornado can get weaker and then stronger many times. This can make it look as if the tornado is skipping over some buildings. In fact, it is always on the ground. But it is doing less damage at weaker times.

A tornado may still be rotating on the ground even if its cloud isn't easily visible.

Strong tornadoes, such as EF4 tornadoes, can have a wedge-shaped appearance.

Wedge Tornadoes

Wedge tornadoes look wide. They often seem broader than they are tall. But the actual size of the funnel might be much smaller. A wedge tornado might simply be kicking up a lot of dust.

Small but Dangerous

A small tornado can be as dangerous as a big one. Wind speed is what matters. A tornado of any size can cause a lot of damage.

Rope tornadoes sometimes spin faster than larger tornadoes.

Rope Tornadoes

Tornadoes can also be skinny. These are called rope tornadoes. Wide tornadoes can turn into rope tornadoes. A tornado can grow stronger the skinnier it gets.

How Tornadoes End

Tornadoes end when they run out of energy. The temperature may get cold. Or the storm clouds might start to break up.

When a tornado breaks up, it is said to be dissipating.

Prediction Problems

Tornadoes are hard to predict. Two storms might have the same temperature. They may have the same wind speed. They could be over the same type of land. One storm forms a tornado. The other one does not. Meteorologists work to try to predict tornadoes.

Some universities, such as Penn State University, have programs that teach students about meteorology and tornado forecasting.

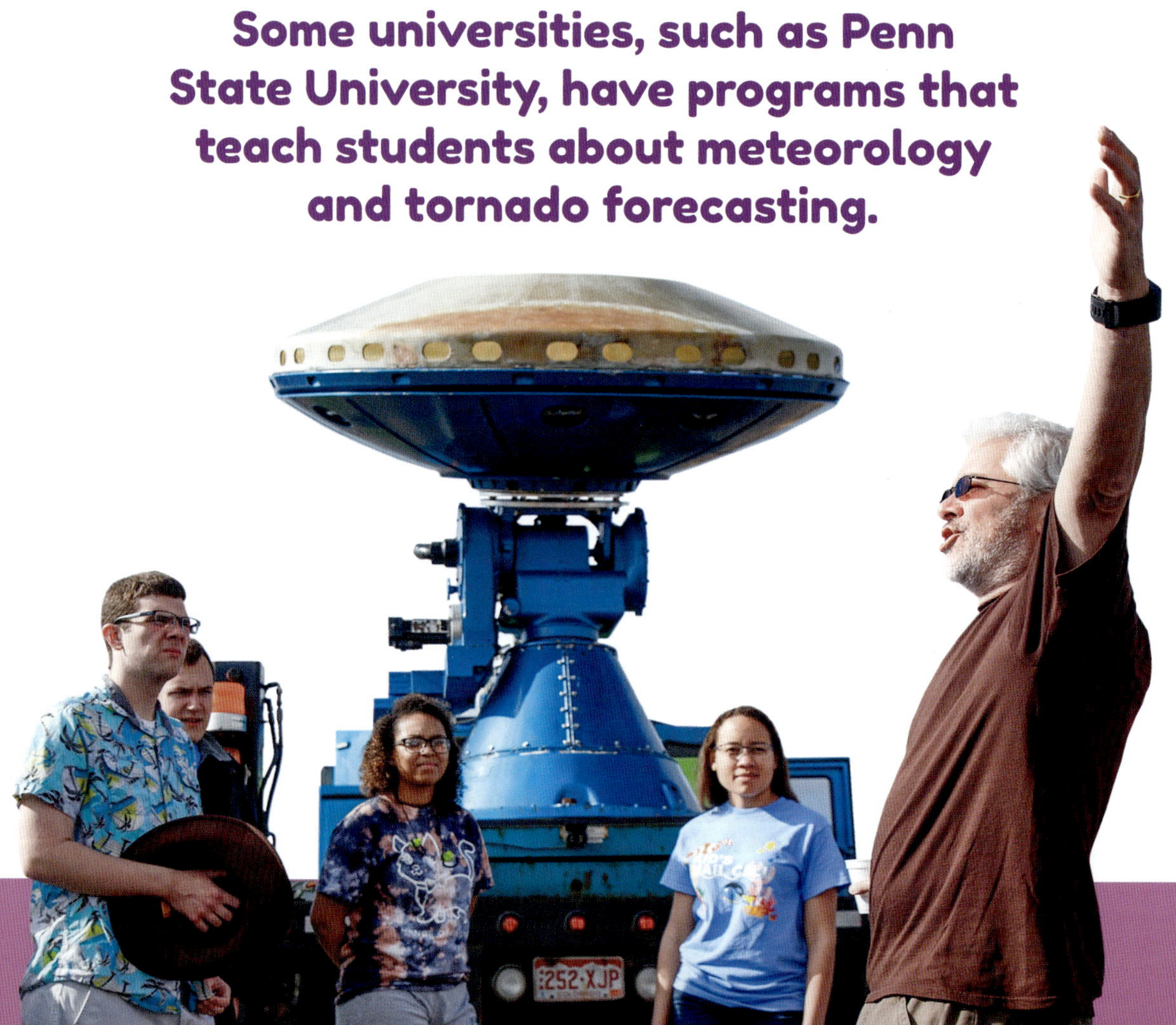

Meteorologists have many tools to help them forecast tornadoes.

Weather Forecasts

A forecast is a prediction of future weather. Forecasts are not certain. Meteorologists use science to figure out the most likely future weather. This science has gotten more accurate over time.

The earliest photo of a tornado was taken in southern South Dakota in 1884.

First US Tornado

The earliest recorded sightings of tornadoes date back about 2,000 years. The first recorded tornado in what is now the United States was in 1643. It happened in what is now Massachusetts.

Banned Word

In the late 1800s, the US government banned the word *tornado* from forecasts. That was because tornadoes were so scary. Forecasts were often wrong. The government did not want to cause panic. The ban lasted until the 1950s.

FUN FACT!

Governor John Winthrop of Massachusetts is credited with making the earliest record of a tornado in 1643.

A photo from the 1890s shows tornado damage in Louisville, Kentucky.

The First Tornado Forecasts

In the 1880s, Sergeant John P. Finley made 15 rules for predicting tornadoes. They included looking for storms with temperature and wind differences. But the word *tornado* was still banned.

John P. Finley was a meteorologist. He lived from 1854 to 1943.

Major Ernest J. Fawbush, *left*, and Captain Robert C. Miller, *right*, were the meteorologists who gave the first tornado forecast.

First Success

In 1948, a tornado hit Moore, Oklahoma. Five days later, a new storm was coming. Two meteorologists compared both storms. The storms were very similar. The scientists said a tornado could come. A tornado hit that day. This was the first correct prediction.

PREDICTING TORNADOES

Researchers launch small weather balloons called windsondes. These tools have sensors that collect data from supercells.

Technology

Weather prediction has come a long way. New technology helps. Computers, radar, aircraft, and satellites all make predictions more accurate.

Measuring Risk

Meteorologists watch how weather systems move. If a storm is likely, they announce an outlook threat level. This lets people know how bad the storm will be. It is a five-point scale. The lowest level is a marginal threat. It goes up to a high threat.

Meteorologists who work for news stations may give tornado threat levels for viewers during weather reports.

Doppler Radar

Doppler radar is a key tool. It was first used in the 1980s. A device sends out microwaves. Those waves bounce off rain and travel back to the device. The way the waves bounce tells how fast a storm is moving. It shows how severe the storm is.

Doppler radar domes can sit on top of communications towers.

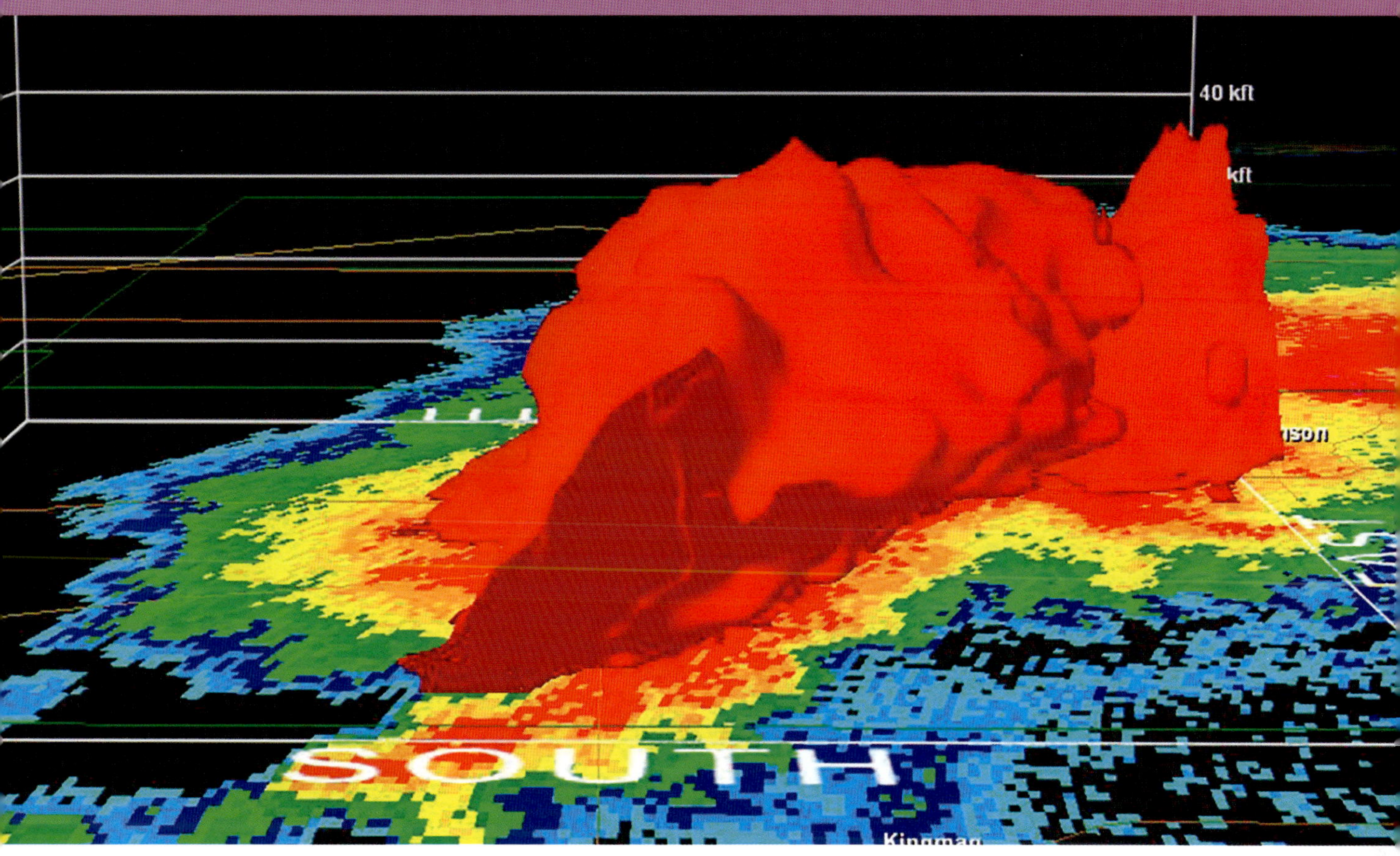

Doppler radar can be used to make 3D models of thunderstorms.

Radar Prediction

Doppler radar shows the direction a storm is moving. It can tell when a storm has winds blowing in different directions. This does not always mean a tornado. But it lets meteorologists warn about a possible tornado.

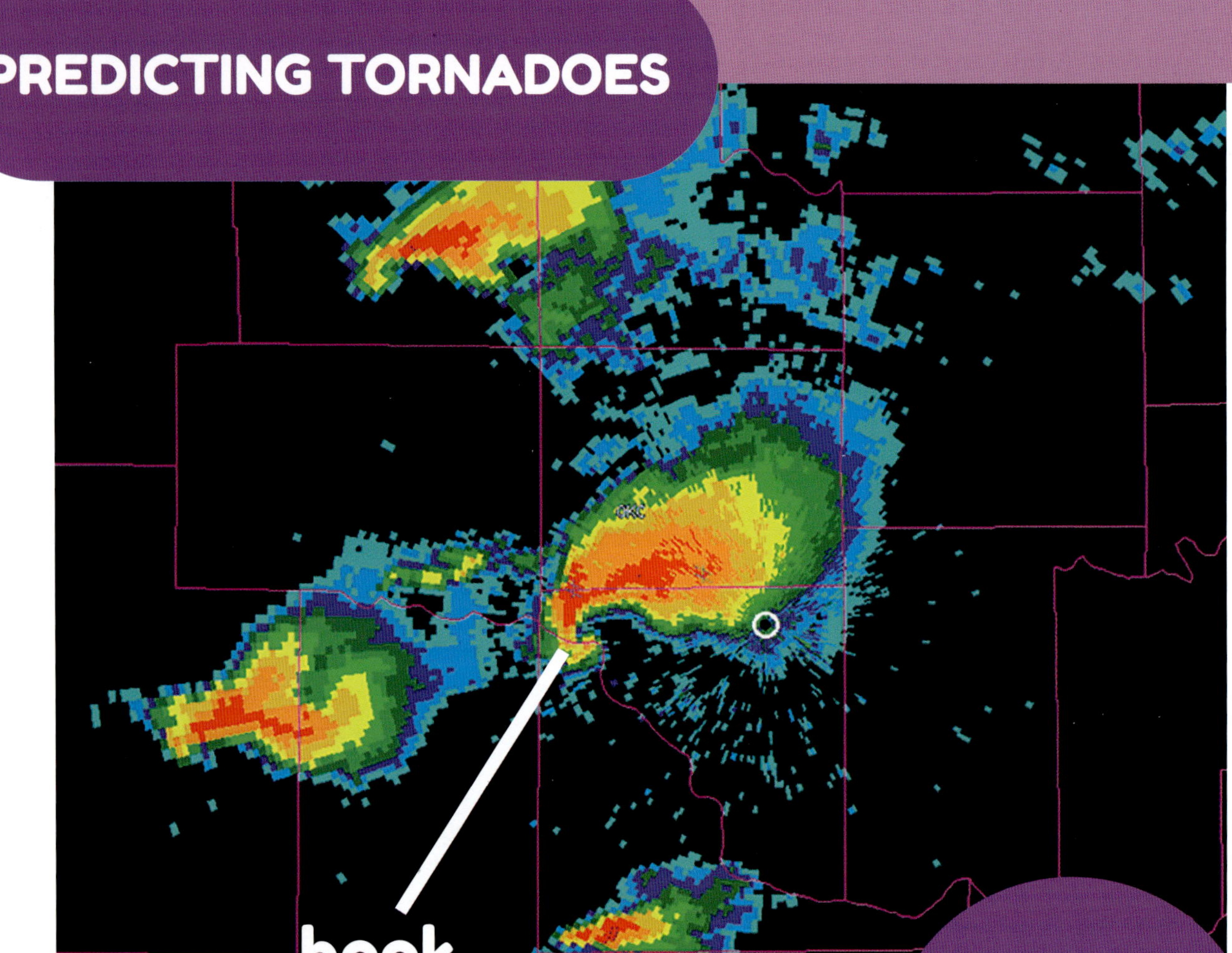

A radar image shows a hook echo, *center.*

Hook Echo

Sometimes a radar image shows a hook shape. The hook is at the back of the storm. The hook means an updraft is sucking rain up into the cloud. That could mean a tornado has formed.

More Warning Time

Scientists can usually warn people about 12 to 15 minutes before a tornado hits. Less than 100 years ago, people had no warning. Even 12 minutes can be enough time to seek shelter.

When people have warning, they can find a safe place to stay during the storm.

Computer Modeling

Scientists are trying to improve tornado prediction. They use data to re-create thunderstorms in computer models. They can study why one storm formed a tornado and another one didn't.

Researchers sometimes use computer models to re-create real tornadoes. This helps researchers study each ingredient needed to make a tornado.

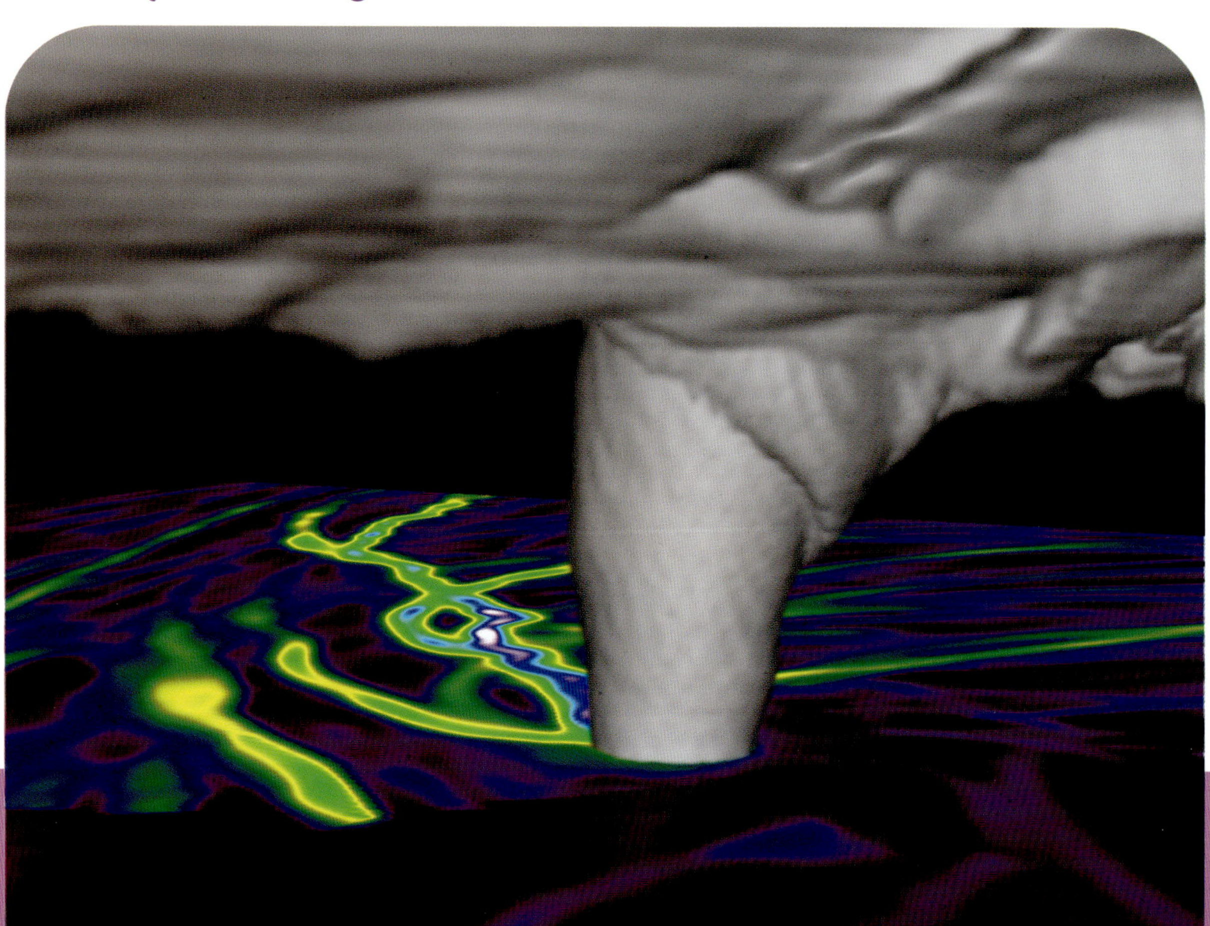

The Ikhana Predator B is one drone that measures wind speeds.

Measuring from the Air

Storms can be thousands of feet in the sky. Devices in the air help scientists get information. Planes carry radar. Drones can measure wind speed and tornado size.

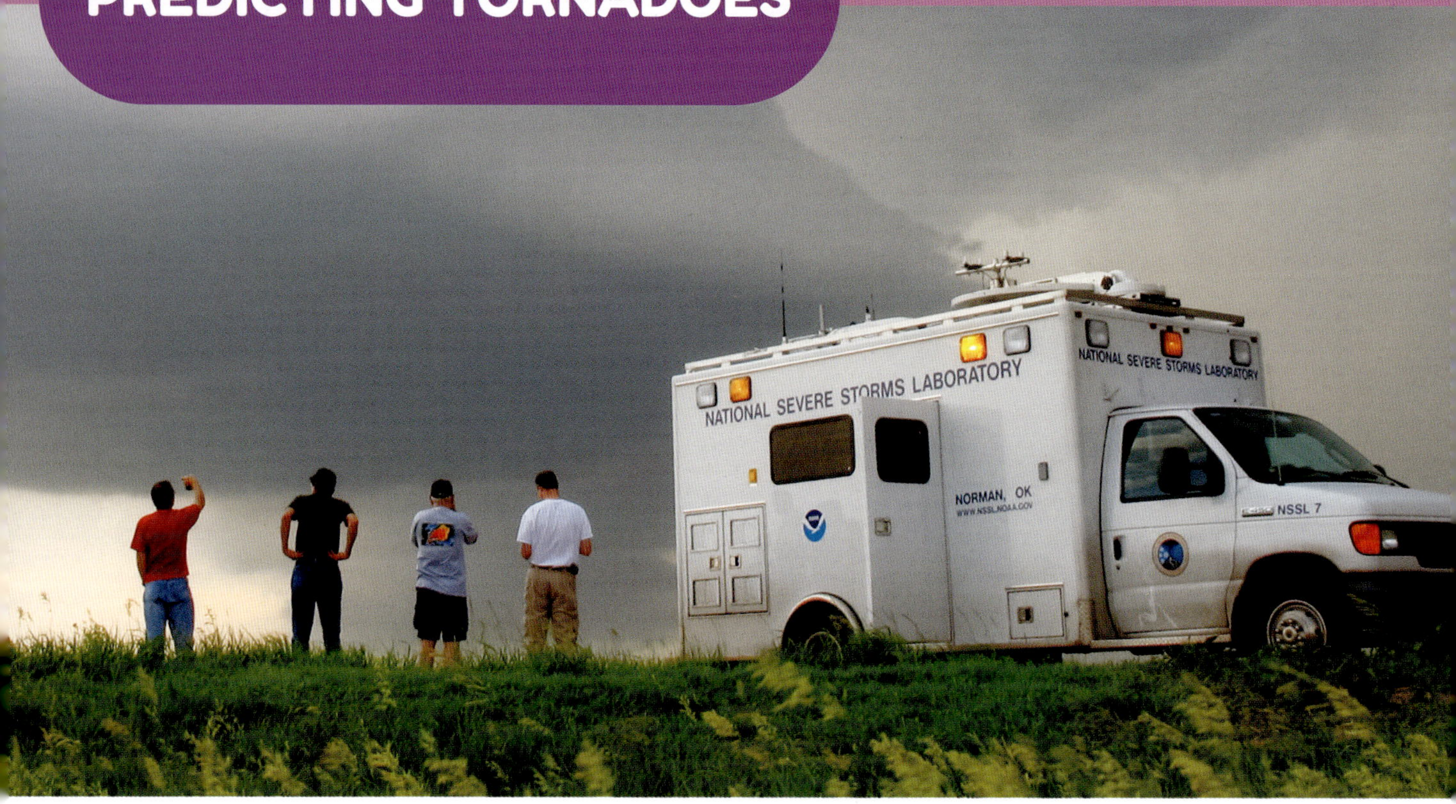

Researchers can drive the VORTEX2 field command vehicle to storms. Its tools help them study storms.

Storm Chasers

Most people should stay away from bad weather. But storm chasers drive to it. Some are scientists. Some are journalists. Some are regular people who love the thrill of a storm.

FUN FACT!

Benjamin Franklin chased a tornado in 1774.

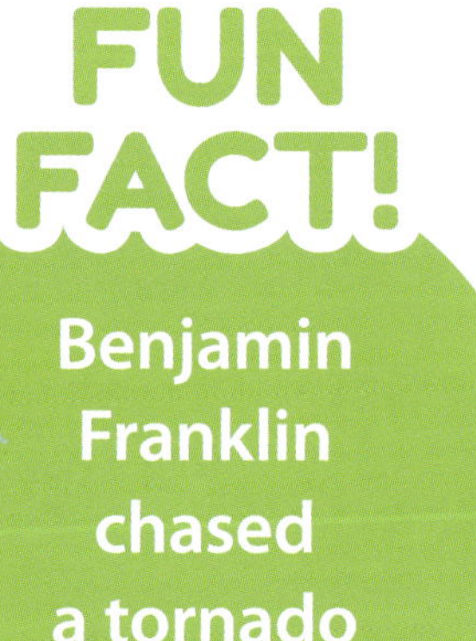

Why Chase?

Some storm chasers are photographers and filmmakers. They take photos and videos of storms. These help scientists understand storms. The images let others see the power of storms. Then people are more likely to take shelter from tornadoes.

It is important that people chasing storms know how to stay a safe distance away.

Chasing Data

Storm-chasing scientists gather data. The data helps create better forecasts. The University of Oklahoma and the National Severe Storms Laboratory teamed up. They started the Tornado Intercept Project in the 1970s. It was the start of storm chasing as a type of research.

Researchers for the Tornado Intercept Project follow a tornado in Oklahoma in 1973.

Doppler on Wheels

Scientists sometimes drive Doppler on Wheels trucks. These are radar trucks. The researchers follow a tornado. The radar gets better data up close.

Storm chasers pay attention to the direction a tornado is traveling. They leave the area if it heads toward them.

Staying Safe

Storm chasers are trained to stay safe. They learn how to keep a safe distance from storms. Too many storm chasers in one area can be a problem. A traffic jam might happen. Then people can't quickly escape from a tornado.

Storm Spotters

The US National Weather Service runs SKYWARN. This is a group of more than 350,000 trained weather spotters. Spotters do not chase storms. They report storm conditions where they are. Their reports help meteorologists understand what is happening in real time.

Storm Chaser Deaths

In 2013, three storm chasers got too close to a storm in El Reno, Oklahoma. A tornado crushed their car. All three died.

The National Weather Service is part of NOAA. Both have offices in Silver Spring, Maryland.

Flying through the Air

The greatest threats during a tornado are projectiles. A projectile is an object thrown through the air with great speed and force. A tornado can hurl a tree branch through a house.

Tornadoes can topple very large trees.

FUN FACT!

Debris can travel more than 100 miles (160 km) during a tornado.

Very powerful tornadoes can lift houses off their foundations.

Damaged Buildings

A tornado can strike a house and fill it with pressure. That pressure can make the roof blow off. It tears the walls away. A whole house can be destroyed in four seconds.

A tornado destroyed mobile homes in Henryville, Indiana, in 2012.

Mobile Homes

Many homes have foundations in the ground. They can resist some wind. But mobile homes are often not attached to the ground. Tornadoes can cause a mobile home to flip over.

Tornado Costs

Tornadoes do an average of $400 million of damage each year in the United States. Some years cost a lot more. In 2023, there were about 1,300 tornadoes. They caused more than $1.2 billion in damage.

The Costliest Tornado

A tornado hit Joplin, Missouri, in 2011. It cost more than $3 billion in damage. It was the costliest tornado in US history.

An EF4 tornado caused great destruction in Rolling Fork, Mississippi, in March 2023.

Injuries and Deaths

Each year, tornadoes hurt 1,500 people in the United States. Flying debris might hit them. They can be hurt if caught in a damaged building. Tornadoes kill about 80 people in the United States yearly. The most common cause of death is head injury.

Those affected by tornadoes may seek comfort from family and friends.

Air Pollution

Tornadoes kick up dust and dirt. The dirty air can affect people with breathing trouble. Tornadoes sweep building materials into the air. One harmful material is asbestos. Asbestos can cause lung damage.

People must be careful to avoid fallen power lines after a tornado.

Electricity

In many communities, electricity runs through lines that hang from poles. Tornadoes can destroy those lines. That might cause a home's power to go out. People can be shocked by fallen power lines. Fallen power lines can also start fires.

Farm Damage

Tornadoes also affect farms. A tornado can easily rip up crops. Farmers insure their crops in case of bad weather. If a tornado destroys a farm, the farmer gets some money back.

Tornadoes can flatten and break crops.

Water Pollution

Tornadoes can cause water pollution. They destroy pipes. That can cause sewage to get into fresh water. Tornadoes can also damage buildings where chemicals are used. Those chemicals can spill into the water. This causes humans and animals to get sick.

Manufacturing facilities contain many chemicals that could pollute the environment.

Fire Tornadoes

Tornadoes can form during wildfires. Hot air reaches up toward the cooler air above. This can create a tornado funnel. Fire tornadoes can cause fires to spread more quickly.

Tornadoes can blow away nests in trees or even the entire tree the nest is in.

Habitat Destruction

High winds can destroy animal homes such as nests. If a tornado passes over a lake or pond, it can suck up fish and other creatures. They may be stranded on dry land.

Tree Damage

Trees take a lot of damage from tornadoes. High winds break off large branches. This wounds a tree. Some wounds can heal. But some kill the tree. Tornadoes can also push a tree over. This kills the tree too.

Trees ripped from their roots cannot take in water and nutrients.

Using Debris for Good

After a tornado, animals are left without homes. Humans can help. They can make piles of branches and leaves. Squirrels and rabbits can hide under the branches.

Fallen branches can create thick cover for small animals.

The Amur honeysuckle is an invasive species in the United States that can take over when tornadoes damage trees.

Invasive Species

Invasive species are animals and plants that harm places where they are not native. They take resources such as food away from native species. When trees are damaged, invasive plants get more sunlight to grow.

Cardinals are among the birds that make nests in shrubs.

Birds

Many birds lose their homes during tornadoes. Those that live in big trees often do not have a place to return to. But some birds benefit from the damage. When trees are killed, more shrubs grow. Birds that thrive in shrubs have a new place to call home.

Hearing Storms

In 2014, hundreds of birds in Tennessee suddenly flew south. A few days later, a massive storm system hit the area. The system created 84 tornadoes. Researchers think the birds could hear the storm coming.

Scars

Tornadoes can leave scars on the ground. Images taken from planes or satellites reveal the scars. They show the path of destruction. The scars can last for years.

A satellite image can show a tornado's path of destruction.

Tornado Watch

Meteorologists know the types of storms that might produce a tornado. If they see one of those storms, they might announce a tornado watch. A watch does not mean a tornado will happen for sure. It tells people to watch their surroundings.

Conditions that cause supercells may be present when a tornado watch is given.

Tornado Warning

A tornado warning means a tornado is in the area. Meteorologists might see one on radar. Or a storm spotter might report one. When a tornado warning happens in the United States, sirens may sound.

FUN FACT!

Tornado sirens were first used in 1970.

Weather radios that can run on batteries are important if the power goes out.

Be Aware

Staying safe from a tornado starts with being aware. If the sky goes dark with clouds, people should check a weather report. The weather radio is one place to hear a report. A computerized voice reads weather announcements 24 hours a day. People should know the nearest shelter. If sirens start, people should seek shelter immediately.

Storm Shelters

A storm shelter is built to survive strong winds. It can be underground. There are also safe rooms. Safe rooms are aboveground. They are built to withstand a tornado.

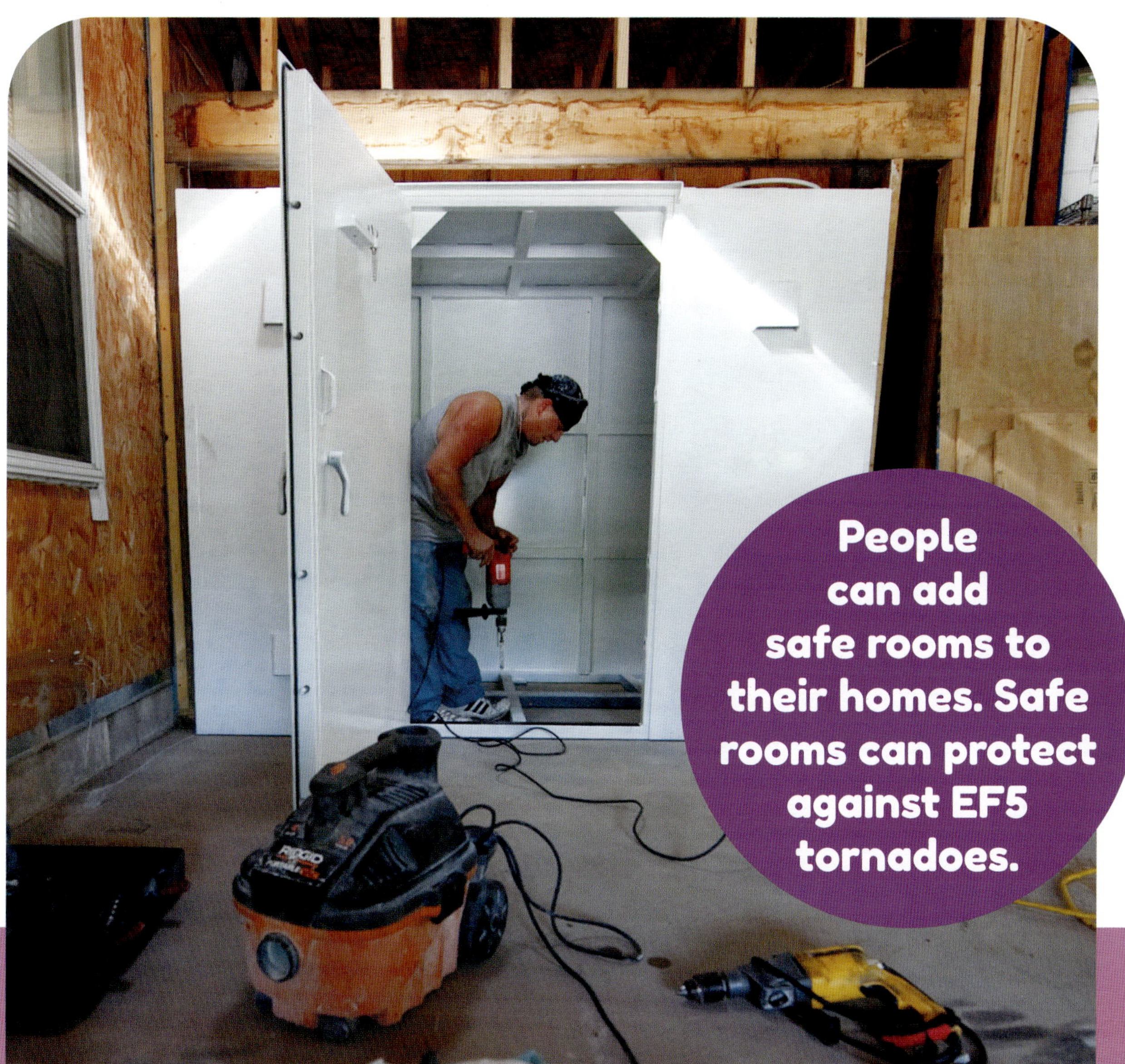

People can add safe rooms to their homes. Safe rooms can protect against EF5 tornadoes.

In a Home

People at home when a tornado strikes should go to the lowest level of the house. They should stay in an interior room. It is important to stay away from windows. The glass can break and hurt people.

A closet without windows is one type of room people can shelter in.

Community shelters can be partially underground. The ground helps provide protection.

Community Shelters

Some cities and businesses have community tornado shelters. Residents can seek shelter there. Such shelters are especially important for people who live in mobile homes.

Storm cellars are sometimes called underground bunkers.

Storm Cellars

Storm cellars are underground shelters. They are holes dug into the ground near a home. Some are lined with plastic or concrete. A tornado can pass over. The people inside stay safe.

Large Buildings

Tornadoes can happen when people are at school, at work, or in other places. People in a large building should go to the lowest level and stay away from windows. They should look for a small room to shelter in.

In some large buildings, public restrooms serve as storm shelters.

Cover Heads

Even in a shelter, people should protect their heads from injury. They can cover their heads with their hands. They could cover themselves with a mattress. Or they can crawl under a heavy table.

Even a thick book can provide some protection for the head.

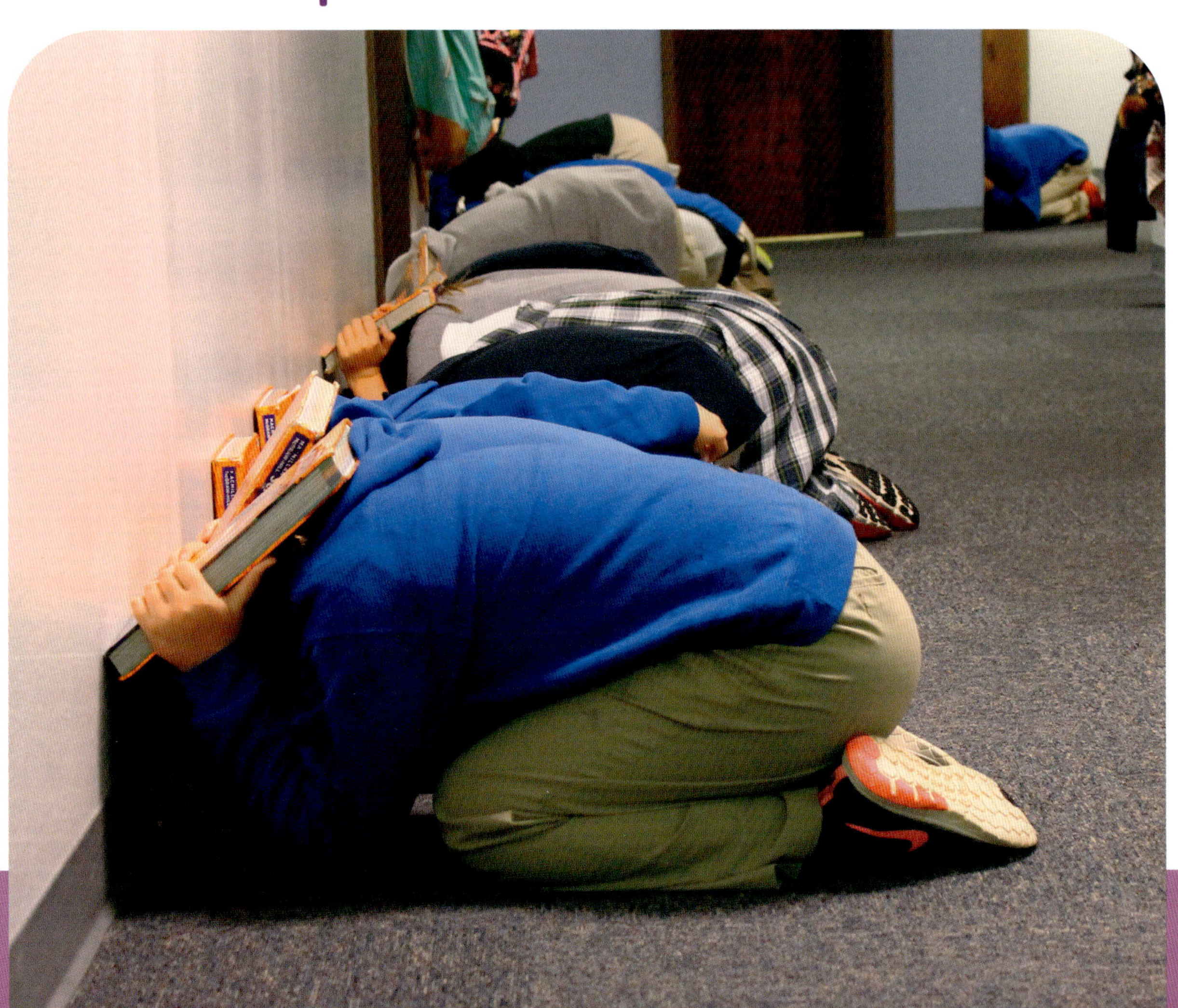

Outdoors

Tornadoes can strike when no shelter is around. If that happens, people should move away from trees. A ditch or ravine can provide some shelter. People should lie as flat as possible. They should cover their heads.

People should park their cars off the road or out of lanes of traffic if a tornado is approaching.

In a Car

People might be in a car and unable to find shelter. They can stay in the car with the seatbelt on. They should try to get lower than the car windows.

Stay Put

Once in a shelter, people should stay put. It is important to follow local weather reports. A thunderstorm can produce many tornadoes. People should stay sheltered until a meteorologist says it is okay to come out.

Where to Seek Shelter

If a house has a basement, that can be a good place to shelter during a tornado. In houses without basements, interior rooms are the safest.

Have a Plan

Every family should have a tornado plan. All family members should know where to go during a tornado warning. They should know where first aid kits and fire extinguishers are.

A first aid kit should include scissors, bandages, and antibiotic ointment, among other items.

During a tornado drill at school, an alarm may sound. People practice their safety plan.

Do Drills

Families, schools, and businesses do tornado drills. During a drill, people pretend a tornado is happening. They seek shelter. They stay in place until the drill is over. Practicing a plan can make a scary moment less frightening.

Emergency Kit

A tornado emergency can last long after a storm has passed. Emergency kits can help. A kit has supplies for a few days. It should have water, food, and a first aid kit. It should also have medications and a change of clothes.

A fireproof safe should be big enough to store all important documents.

Important Documents

If a house is destroyed, important documents might be lost. Birth certificates, pet identification, and medical information should be kept in a safe place. A waterproof and fireproof safe is one option. So is keeping copies of all important documents in an emergency supply kit.

Having a plan in place can save a pet's life if a tornado comes.

Keeping Pets Safe

Pets should be part of an emergency plan. An emergency kit should have their medications, identification, and food. It is important to know where pets usually hide. That way, owners can find them quickly. Owners can then bring their pets to safety.

Trapped

Some people get trapped in buildings or under debris. They should make noise to help rescue workers find them. They can call for help or send a text. Some people keep a whistle in their emergency kit. They blow the whistle. Rescuers can hear them.

Search and rescue dogs sometimes help find people trapped after tornadoes.

Cleaning Up

Tornado cleanup is often dangerous. Damaged buildings could collapse. There could be gas leaks. Adults should handle the cleanup.

Some non-profit organizations send volunteers to help communities clean up after a tornado.

Wearing gloves and other protective clothing is important for anyone cleaning up.

Damaged Buildings

Only trained professionals should go near damaged buildings. Even the outside of a building can be dangerous. Tornadoes often break windows. Shards of glass could be on the ground.

The Red Cross

The Red Cross is an organization that helps after a disaster. Tornadoes can damage or destroy homes. The Red Cross creates shelters for people to stay in. It helps with cleanup.

Gas Leaks

People use gas for heating and cooking. Tornadoes can break pipes and cause gas to leak. A rotten egg smell might mean gas is leaking. Any spark or flame could cause an explosion. People should leave and call the gas company.

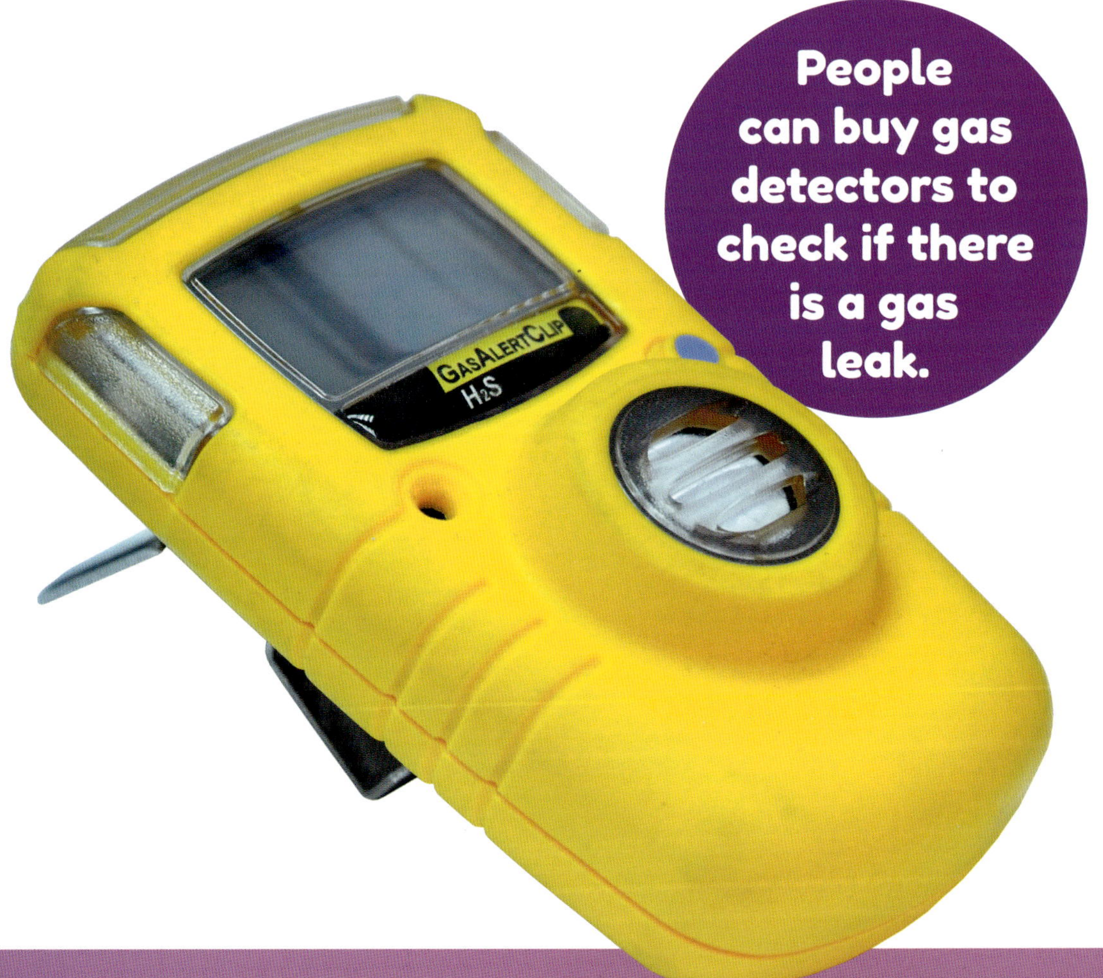

People can buy gas detectors to check if there is a gas leak.

It is important to have plenty of clean water to drink when cleaning up after a tornado.

Food and Water Safety

Refrigerators do not work when the power is out. Food can go bad. Sometimes storms cause water to become unclean with bacteria or chemicals. If that happens, people should drink only bottled water. They should use it to make food and brush their teeth.

A desert is a place with a dry climate. It is generally dry year-round.

Weather vs. Climate

Weather measures what happens on a given day. It might be sunny or rainy. It might be cold or hot. Climate measures what happens over time. It considers the weather in a place across many years.

Climate Change

Since the 1800s, humans have been causing changes in the climate. Fossil fuels such as coal release greenhouse gases. Those gases trap heat in the atmosphere. This causes the planet to warm.

Many vehicles burn a type of fossil fuel called gasoline.

Climate Change Effects

Rising temperatures change weather patterns. Storms are becoming more severe. Climate change affects how much rain falls. Some places see more flooding. Others have droughts.

Flooding causes widespread damage to property.

A photograph taken in 1959 shows a tornado in Kansas.

Recording Tornadoes

It is hard to know how climate change has affected tornadoes. Researchers have fewer than 100 years of records. This is because of the past ban on the word *tornado*. There are other reasons too.

A meteorologist storm chaser monitors a storm in a Doppler on Wheels vehicle.

Seeing More Tornadoes

Doppler radar is used more today than in the past. Storm chasing is also more popular. So more tornadoes are now spotted. That does not mean more are happening. It just means people are better at spotting them.

Short-Lived

Tornadoes are relatively small. They do not last long. That makes it hard for scientists to get enough information about ways they are changing.

Even powerful tornadoes, such as EF4s, often last for less than an hour.

Looking at Ingredients

Scientists look at the key ingredients. They see how climate change affects the ingredients. Scientists study strong thunderstorms. They study wind shear.

Radar can help scientists find thunderstorms to study.

Researchers studying tornadoes have portable weather stations that measure wind data, temperature, and more.

Air Conditions

Climate change causes more warm, humid air. That creates more air instability. Unstable air can cause severe storms to become more frequent. Severe storms can create tornadoes.

Wind Shear

The jet stream powers wind shear. Strong winds in the atmosphere make up the jet stream. It is strongest when there is a large temperature difference between warm oceans and cold poles. But climate change is causing the poles to warm. There is less of a temperature difference. The jet stream is weaker.

Researchers measure wind speed using anemometers.

More and Less

Climate change could make storms stronger than ever. But it might not lead to more or stronger tornadoes. Lack of wind shear could mean no updraft. No updraft means no tornadoes.

Climate change leads to more lightning strikes.

More Clusters

The number of tornadoes each year is similar. But how they appear is changing. There are fewer days with just one tornado during tornado season. But there are more days with more than 30 tornadoes. That means tornadoes are now more likely to come in clusters.

Clusters cause almost 80 percent of tornado deaths in the eastern two-thirds of the United States.

In March 2023, several tornadoes tore through Mississippi and Alabama.

Changing Geography

Tornadoes are also happening more often in the US East and South. There are more tornadoes in Mississippi and Alabama. There are also more in Illinois and Indiana. The number in the Great Plains is going down.

Houses not built with tornadoes in mind may offer less protection to people sheltering there.

Geography Matters

The South has a higher population than the Great Plains. Many houses in the South are not built with places to take shelter. So tornadoes could harm more people than before.

Different Seasons

In the United States, tornadoes usually happen in spring and summer. That could shift with climate change. More tornadoes could happen in autumn and winter.

A tornado tore through Madison, Tennessee, on December 9, 2023.

Touchdown in Missouri

On March 18, 1925, a storm was brewing. At 1:00 p.m., a tornado formed in Ellington, Missouri. There was no tornado warning system. Residents did not know it was coming. It traveled northeast. It killed 11 people in the state.

The Tri-State Tornado passed through a largely rural part of southeastern Missouri, including Biehle.

West Frankfort, Illinois, was badly damaged during the Tri-State Tornado.

Moving to Illinois

The EF5 tornado moved to Illinois. Its winds blew up to 300 miles per hour (480 kmh). It grew to one mile (1.6 km) wide.

Caught Off Guard

The tornado still surprised people. It killed more than 600 people in Illinois. It greatly damaged several Illinois towns.

Betty Moroni of De Soto, Illinois, survived the Tri-State Tornado. She was seven years old when it struck.

The tornado's strong winds flipped houses.

Arriving in Indiana

The tornado raced at more than 70 miles per hour (110 kmh). Most tornadoes travel at about 10 to 20 miles per hour (16 to 32 kmh). The Tri-State Tornado reached Indiana at 4:00 p.m.

People searched the wreckage of Griffin, Indiana, after the tornado.

More Destruction

The tornado killed 71 people in Indiana. In Griffin, Indiana, every single building was destroyed. The storm ended at 4:30 p.m.

A Long Track

The Tri-State Tornado was one of the longest-lived tornadoes in US history. It was on the ground for 219 miles (352 km). The tornado lasted for 3.5 hours.

Major Damage

The Tri-State Tornado did major damage. It destroyed more than 15,000 homes. Some soldiers compared the destruction to war damage.

Tri-State Tornado Path

The Tri-State Tornado damaged or destroyed towns in Missouri, Illinois, and Indiana.

Deadly Event

No other US tornado has caused as many deaths. It killed 695 people. It injured more than 2,000 people.

In De Soto, Illinois, 33 children died when the tornado struck a school.

The tornado destroyed about half of Princeton, Indiana.

More than One Tornado?

Some scientists have wondered if the Tri-State Tornado was more than one tornado. No other tornado like it has ever been seen. But the path of damage was almost continuous. This means it was likely a single tornado.

Bangladesh Tornado

The deadliest tornado ever happened in Bangladesh. In 1989, the tornado hit two towns. More than 1,300 people died. It injured 12,000 people.

The First Storms

On April 25, 2011, storms raced across the South. One tornado killed four people near Little Rock, Arkansas. This was the start of a four-day outbreak.

A satellite image showed clouds forming on April 25, 2011.

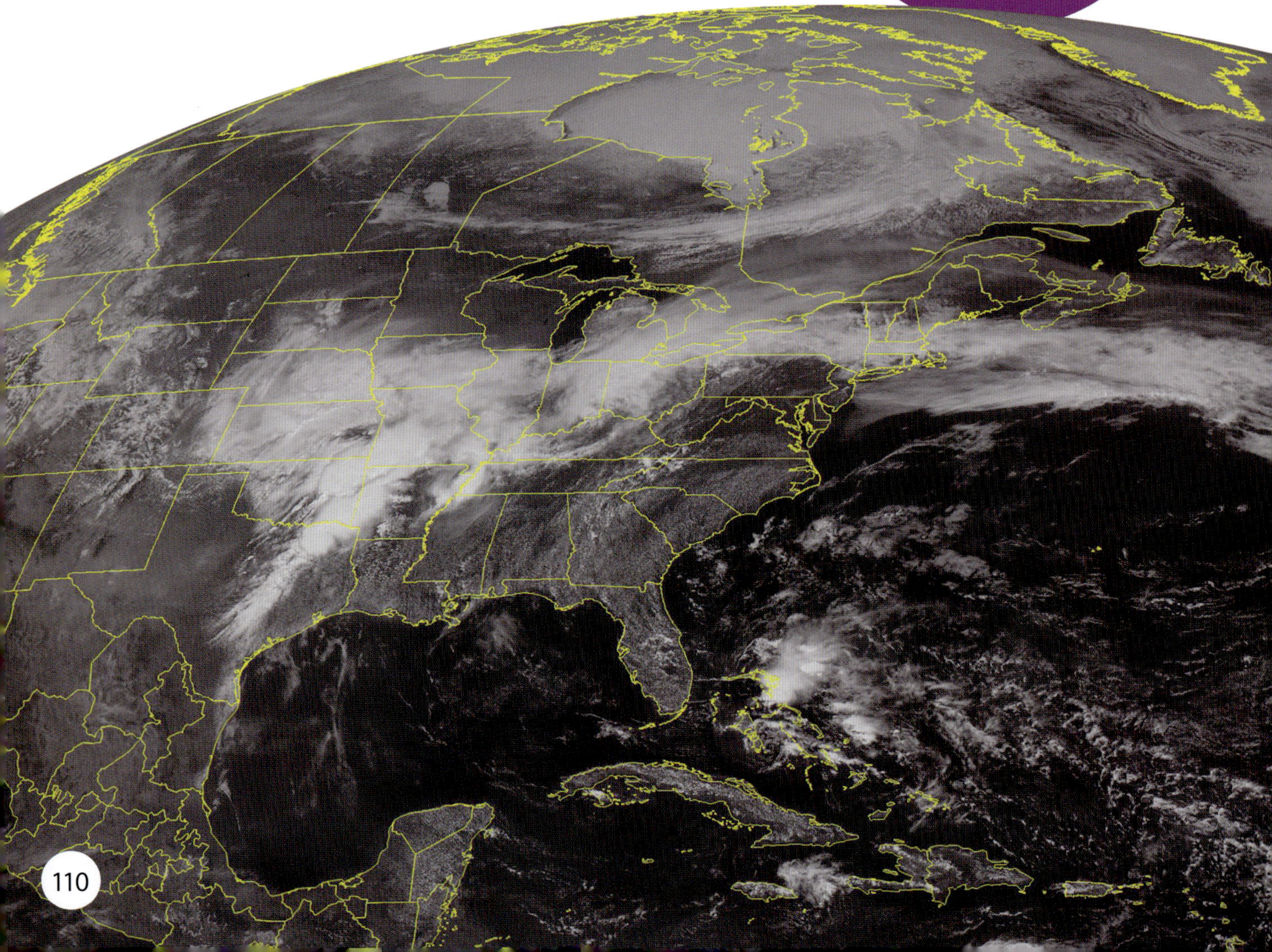

People in Vilonia, Arkansas, picked up after an April 25 tornado.

The Second Night

April 26 brought more tornadoes. An EF3 tornado hit Fort Campbell, Kentucky. The wind blew at 140 miles per hour (225 kmh).

Many Tornadoes

April 2011 saw 817 tornadoes in the United States. That was the most tornadoes in a single month.

By April 27, storm clouds were dense. The jet stream blew into the clouds, causing storms to rotate.

Predicting the Next Storms

Meteorologists knew April 27 would be a dangerous day. They said that morning that strong tornadoes were possible. They were right.

EF5

The strongest tornado of the outbreak hit Hackleburg, Alabama, at 3:05 p.m. It lasted more than two hours. It traveled 132 miles (212 km). The tornado's path led it through other cities, including Phil Campbell, Alabama. The EF5 tornado's winds reached 210 miles per hour (338 kmh). It threw cars 150 to 200 yards (140 to 180 m). It killed nearly 100 people.

Hackleburg High School seniors graduated outside their destroyed school on May 27, 2011. This was exactly one month after the tornado hit.

Photojournalist Dusty Compton captured a photo of the tornado in Tuscaloosa.

Tuscaloosa and Birmingham

At 4:43 p.m., an EF4 tornado also formed in Alabama. It traveled 80 miles (129 km). It hit both Tuscaloosa and Birmingham in Alabama. The tornado killed 65 people. It injured 1,500.

An End at Last

The storms continued through April 28. The system finally blew out over the Atlantic Ocean. The storms were over.

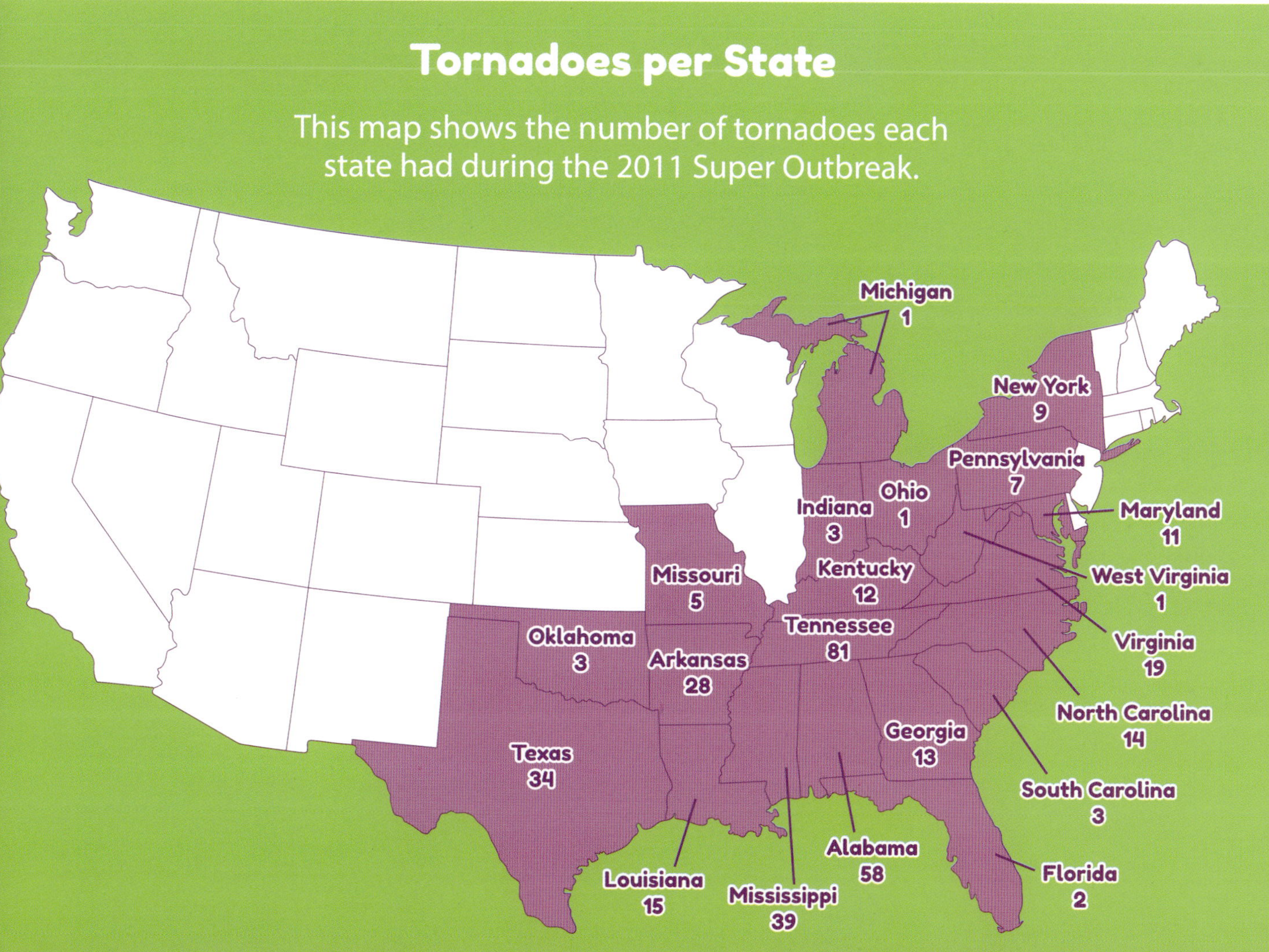

Effects

The 2011 outbreak produced three EF5s. It caused about $12 billion in damage. It was the most expensive outbreak in US history. In Alabama alone, tornadoes created 1,200 miles (1,930 km) of tracks.

FUN FACT!

Less than 1 percent of tornadoes are EF4 or EF5.

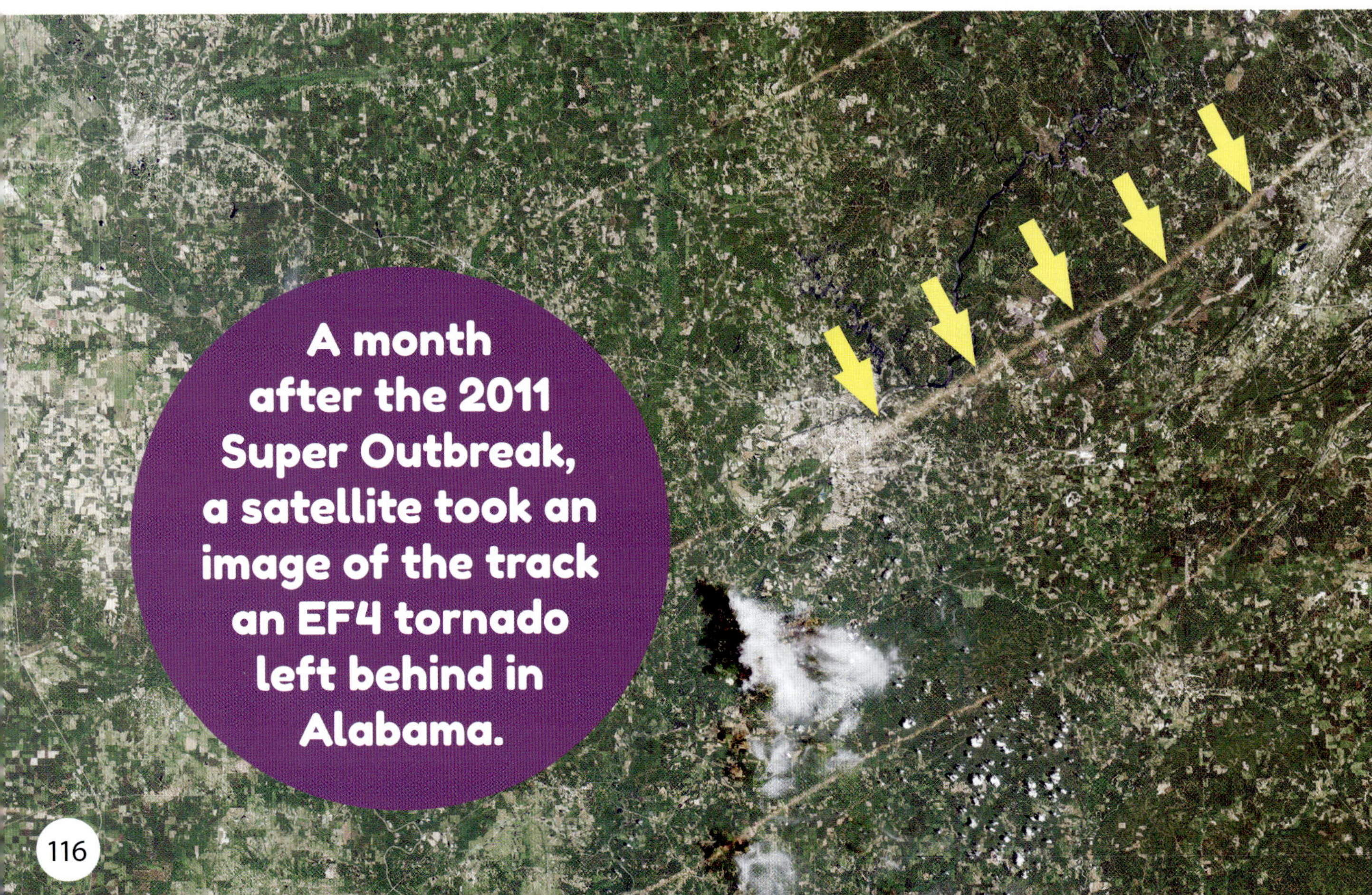

A month after the 2011 Super Outbreak, a satellite took an image of the track an EF4 tornado left behind in Alabama.

First Lady Michelle Obama visited a relief center in Holt, Alabama, on April 29.

Injuries and Deaths

The outbreak had more than 360 tornadoes. More than 320 people died. Of those, 253 deaths were in Alabama. More than 3,000 people were injured.

Alabama on April 27

On April 27, 2011, 62 tornadoes hit Alabama during an 18-hour span. The first was at 4:01 a.m. The last ended at 9:58 p.m.

THE 1999 AND 2013 MOORE TORNADOES

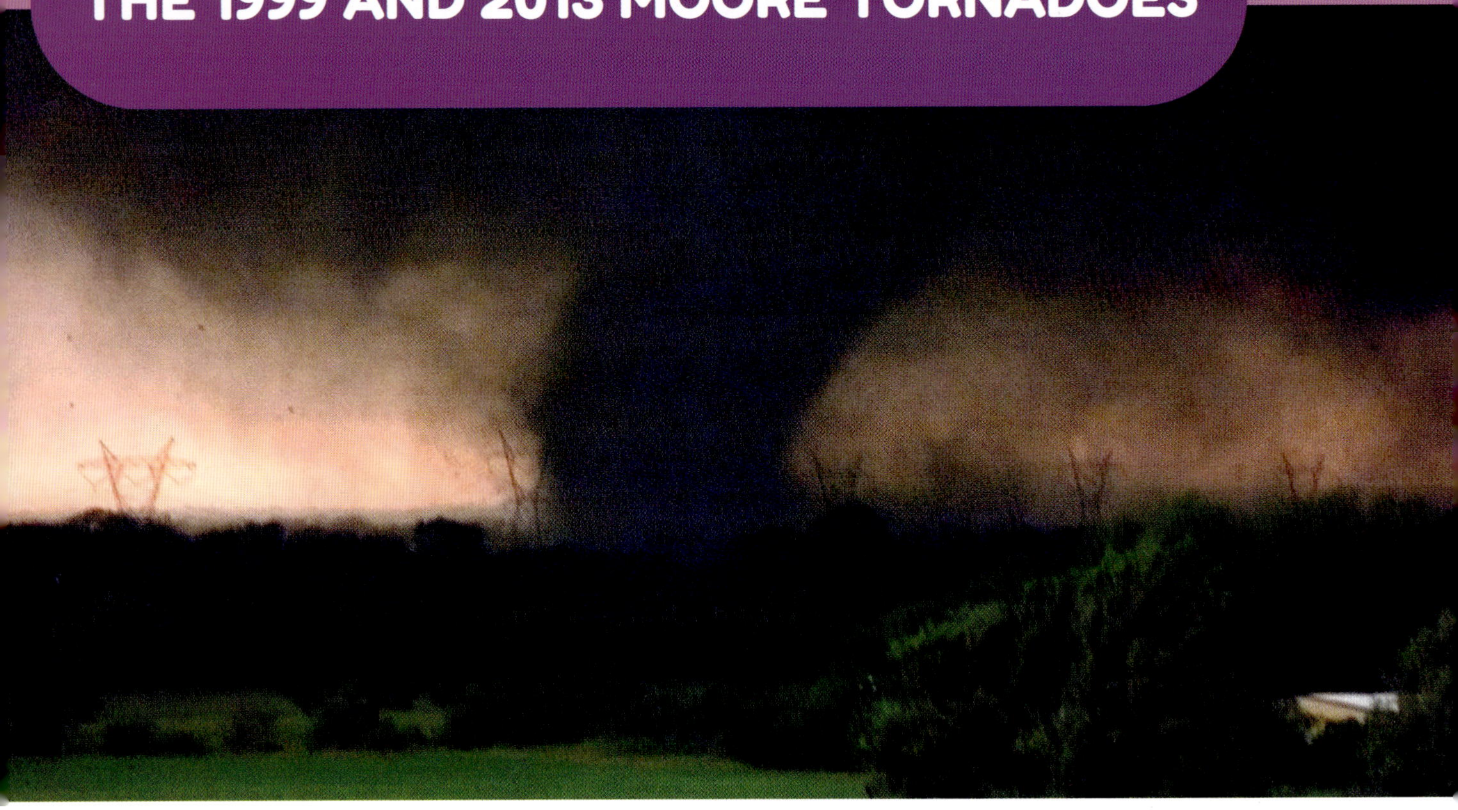

The 1999 Moore tornado was about 0.5 miles (0.8 km) wide.

1999 Outbreak

It was just before 4:00 p.m. on May 3, 1999. Storms started moving through Oklahoma. The first tornado struck at 4:51 p.m. It did not cause damage. But others would.

FUN FACT!

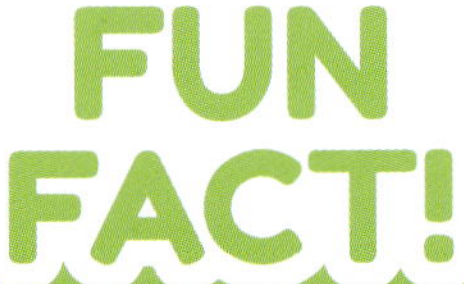

Tornadoes are also known as twisters.

The 1999 Moore Tornado

At 6:23 p.m., an EF5 tornado struck Moore, Oklahoma. It had wind speeds of more than 300 miles per hour (480 kmh). It lasted 1.5 hours. It traveled 38 miles (61 km).

The tornado damaged large stretches of homes.

Effects

The Moore tornado killed 46 people. About 800 were injured. Around 2,500 homes were damaged. It destroyed nearly 1,800 homes. The tornado did more than $1 billion in damage.

Postal carriers began delivering mail to neighborhoods a few days after the tornado as people began the cleanup.

Overpasses often create narrow spots in the road. If the road gets blocked here, other cars cannot pass by.

Rethinking Cover

Several people died beneath overpasses during the EF5 tornado. Overpasses are bridges that allow roads to pass over other roads. People had stopped under them to shelter from the tornado. This blocked traffic. Some people couldn't escape the oncoming tornado. So experts now say people should never stop under an overpass during a tornado.

A tornado formed near Viola, Kansas, on May 19, 2013.

2013 Outbreak

A tornado struck Ness County, Kansas, on May 18, 2013. It was the start of a tornado outbreak that would last until May 20. On May 19, tornadoes hit Kansas, Nebraska, Iowa, Missouri, and Oklahoma.

Hit Again

On May 20, at 2:56 p.m., an EF5 tornado again hit Moore. It lasted 40 minutes. The tornado's track was 14 miles (23 km) long. The tornado was 1.1 miles (1.8 km) wide at times.

A photographer captured the EF5's approach to Moore.

Effects

The EF5 did $2 billion worth of damage. Twenty-four people died. This included seven children in an elementary school. Another school was also destroyed. No one died there.

The Same Place Twice

Some areas are more prone to tornadoes than others. A myth is that tornadoes cannot strike the same place twice. But the Moore tornadoes did.

Unlucky Spot

An F4 tornado struck Moore in 2003. That means three massive tornadoes struck in the same place across 14 years.

Photographers took photos of the destruction, *top*, and rebuilding of Moore one year later, *bottom*.

GLOSSARY

debris
The remains of something broken or destroyed.

drone
A small, remote-controlled aircraft.

geography
The physical features of the planet's surface.

instability
Being unsteady.

insure
To get insurance, which is when someone pays small fees to a company that agrees to pay a larger fee if something is lost or destroyed.

microwave
A wave of energy that can bounce off an object.

pollution
Materials that enter nature as the result of human activity and are harmful to the environment.

pressure
A force that presses against something.

rotate
To spin.

sewage
Wastewater from homes and buildings, such as from sinks or toilets.

shrub
A small plant such as a bush.

siren
A warning device that creates a loud sound.

TO LEARN MORE

More Books to Read

Buckey, A. W. *Weather.* Abdo, 2025.

Challoner, Jack. *Hurricane & Tornado*. DK, 2021.

Gieseke, Tyler. *Weather Cycles*. Abdo, 2023.

Online Resources

To learn more about tornadoes, please visit **abdobooklinks.com** or scan this QR code. These links are routinely monitored and updated to provide the most current information available.

INDEX

PHOTO CREDITS

Cover Photos: Benjamin Simeneta/Shutterstock Images, front (house); Minerva Studio/Shutterstock Images, front (tornado); Shutterstock Images, front (students), back

Interior Photos: Joni Hanebutt/Shutterstock Images, 1; Shutterstock Images, 3, 6 (bottom), 8, 15 (bottom), 17 (top), 22, 25, 27 (bottom), 48, 53 (top), 62, 63, 64 (bottom), 74, 76, 77, 78, 80, 86, 87, 89, 90, 93, 96, 97, 109 (bottom), 111 (bottom), 117 (bottom), 124; Minerva Studio/Shutterstock Images, 4, 5, 19; Eugene R. Thieszen/Shutterstock Images, 6 (top); Emily Summars/University of Oklahoma Cooperative Institute for Mesoscale Meteorological Studies/NOAA, 7; NOAA/NASA/Climate.gov, 9; Danita Delimont/Shutterstock Images, 10; John D. Sirlin/Shutterstock Images, 11, 20; Munir uz Zaman/AFP/Getty Images, 12; Menno van der Haven/Shutterstock Images, 13; Guillermo Guerao Serra/Shutterstock Images, 14; Laura Hedien/Shutterstock Images, 15 (top); Blue Ring Media/Shutterstock Images, 16 (clouds); Red Line Editorial, 16 (cylinder and right tornado), 21, 107, 115; Victoria Sergeeva/Shutterstock Images, 16 (middle tornado); Jim H. Walling/Shutterstock Images, 17 (bottom); Gary Hincks/Science Source, 18; NOAA, 23, 29, 32, 39, 40, 44, 47, 108; National Weather Service, 24, 46; Melanie Metz/Shutterstock Images, 26, 67; Roger Hill/Science Source, 27 (top); David Schliepp/Shutterstock Images, 28; Drew Angerer/Getty Images News/Getty Images, 30, 45, 92, 94; Dennis MacDonald/Shutterstock Images, 31; George Barker/Library of Congress, 33; Library of Congress, 34; Cynthia Fawbush Goff/US Air Force/National Weather Service, 35; Christiaan Patterson/OU CIMMS/NOAA, 36; Matt McClain/The Washington Post/Getty Images, 37; Alan B. Schroeder/Shutterstock Images, 38; iStockphoto, 41, 71, 121; Leigh Orf/MDPI, 42; Ken Ulbrich/NASA, 43; Rashad Ashur/Shutterstock Images, 49 (top); Mark Van Scyoc/Shutterstock Images, 49 (bottom); Harrison Jeffs/Shutterstock Images, 50; Benjamin Simeneta/Shutterstock Images, 51; Alexey Stiop/Shutterstock Images, 52, 56, 85 (top); Chandan Khanna/AFP/Getty Images, 53 (bottom); Scott Olson/Getty Images News/Getty Images, 54, 82, 83; Ryan McGinnis/Moment/Getty Images, 55; Matthew G. Eddy/Shutterstock Images, 57; Sean Rayford/Getty Images News/Getty Images, 58; David Gray/Getty Images News/Getty Images, 59; David Young/dpa/picture alliance/Getty Images, 60; John Tlumacki/Boston Globe/Getty Images, 61; Mike Truchon/Shutterstock Images, 64 (top); Lauren Dauphin/USGS/NASA, 65; Cammie Czuchnicki/Shutterstock Images, 66; Guy J. Sagi/Shutterstock Images, 68; Julie Denesha/Getty Images News/Getty Images, 69; Jason Finn/Shutterstock Images, 70; Richard A. McMillin/Shutterstock Images, 72; Lorenza Ochoa/Shutterstock Images, 73; Ruud Morijn Photographer/Shutterstock Images, 75; Marlin Levison/Star Tribune/Getty Images, 79; Kevin Brine/Shutterstock Images, 81; Gino Santa Maria/Shutterstock Images, 84; Jane Kelly/Shutterstock Images, 85 (bottom); Kenneth Keifer/Shutterstock Images, 88; Bert Hoyt/Three Lions/Hulton Archive/Getty Images, 91; Seth Herald/AFP/Getty Images, 95; Alexander Jung/Shutterstock Images, 98; Fatih Aktas/Anadolu Agency/Getty Images, 99; Lokman Vural Elibol/Anadolu Agency/Getty Images, 100; Jon Cherry/Getty Images News/Getty Images, 101; larrybraunphotography.com/Moment/Getty, 102; Bettmann/Getty Images, 103; Byron Hetzler/The Southern Illinoisan/AP Images, 104; AP Images, 105; Topical Press Agency/Hulton Archive/Getty Images, 106; Archive Farms/Hulton Archive/Getty Images, 109 (top); Weather Underground/AP Images, 110; Mark Humphrey/AP Images, 111 (top); GOES Project Science Team/NASA, 112; Butch Dill/AP Images, 113; Dusty Compton/Tuscaloosa News/AP Images, 114; Jesse Allen/USGS/NASA, 116; Saul Loeb/AFP/Getty Images, 117 (top); Paul Hellstern/The Daily Oklahoman/AP Images, 118; Tannen Maury/AFP/Getty Images, 119; Hector Mata/AFP/Getty Images, 120; Travis Heying/The Wichita Eagle/AP Images, 122; Vincent Deligny/AFP/Getty Images, 123; Tom Pennington/Getty Images News/Getty Images, 125 (top); Joe Jaedle/Getty Images News/Getty Images, 125 (bottom)